bup

German Bibliothek
-CIP-Einheitsaufnahme-

Thomas Degenhardt
Why Tesla?
ISBN: 978-3-95562-980-9
Copyright by bremen university press
Bremen, Deutschland
Edition 1, 24. September 2023
Version 1.0
Printed in EU, UK, USA, JP, AUS
bup@bremenuniversitypress.com
www.bremenuniversitypress.com

WHY TESLA?

Why Tesla?

Tesla.

Fifteen years ago, at best another niche producer, by nerds for nerds. Ridiculed, sometimes marveled at, heavily influenced by the eccentric appearances of his face, Elon Musk. That didn't always help. Electric cars at that. Unecological, unreliable, lame, ugly, dangerous, only something for the city, insanely expensive, manufactured using child labor in some mine or other, the entire Internet knew that! Again, someone who wanted to make a small fortune and burned his great fortune to do so. Sigh with pity, buy the next diesel. Topic checked off.

But those who were smart, and there were only a few, bought Tesla shares for 17 dollars each. And later built houses from the profits.

Before you could even rub your eyes, Tesla had managed to create an iconic brand in no time at all that was in no way inferior to Apple's, and incidentally revolutionize the automotive world. Long-range, rechargeable anywhere, desirable sedans faster than supercars, all familiar prejudices shoved down the chimney. Driving the established car manufacturers before them. Until today, by the way. The gap feels like it's getting wider rather than narrower.

How is something like this possible? How can the lead of the major established manufacturers, some of which have existed for 100 years, be overturned with a single new vehicle generation from a newcomer without any experience in vehicle construction?

Many of them had a taste of Tesla in the beginning, but none of them stayed. That was probably a misjudgement.

How did it come about?

Tesla, Inc, originally known as Tesla Motors, was founded in 2003 by Martin Eberhard and Marc Tarpenning in Palo Alto, California. The idea behind Tesla was to prove that electric vehicles would not have to compromise or sacrifice range, performance and functionality.

In 2004, Elon Musk joined the company and became the main investor. Although he was not initially one of the founders, Musk made a significant contribution to the company's strategy and vision and was later described as a co-founder. Today, he is regarded - certainly with good reason - as "Mr. Tesla".

Tesla's first vehicle, the Roadster, was launched in 2008. It was a sports car based on the Lotus Elise chassis, equipped with an all-electric powertrain. With a range of about 245 miles per charge, the Roadster set a new standard for electric vehicles.

In 2012, Tesla introduced the Model S, an all-electric luxury sedan that quickly gained recognition for its range, performance and design. The Model S could travel over 300 miles on a single charge and was highly praised by critics. With the success of the Model S, Tesla established itself as a serious competitor in the automotive industry.

In 2015, Tesla introduced the Model X, an all-electric SUV. Despite initial production challenges due to its complex "falcon wing" doors, the Model X was well received, especially in the US.

The real game-changer, however, came in 2017 with the launch of the Model 3, a more affordable electric vehicle for the mass market. Although there were initial production delays, the Model 3 quickly became a bestseller.

Tesla also expanded globally, with factories in Shanghai, China and additional factories in Europe and parts of Asia.

In addition to electric vehicles, Tesla has also developed other products and services. These include the Powerwall home storage, the Powerpack for commercial applications and the SolarRoof products.

The company has also made progress in the field of autonomous driving with the aim of developing fully self-driving cars. A long road, the core activities of which will also be highlighted here.

The founders of Tesla

Tesla is known today as one of the leading companies in the field of electric vehicles. Like any great company, Tesla has its own unique founding story.

The little-known fact is that Tesla Motors (as it was originally called) was founded by Martin Eberhard and

Marc Tarpenning in 2003. Eberhard and Tarpenning initially funded the fledgling company themselves and were the driving forces behind the development of Tesla's first vehicle, the Tesla Roadster. Although not one of Tesla's original founders, Elon Musk played an important role in the company's early days. He joined Tesla shortly after its founding and quickly became its largest investor. Musk took active roles in company management and product planning, and although he was not originally considered a founder, he was later even recognized as a co-founder of Tesla after some legal wrangling. Musk's vision, ambition and financial support helped guide the company through its critical early years.

Over time, many other talented individuals joined Tesla and have been instrumental in the company's growth and success. While they may not hold the title of "founder," their contributions to Tesla's development and expansion cannot be overlooked.

Tesla's success is the result of the vision, determination and innovation of a group of individuals who believed in the possibility and necessity of pure electric vehicles. While Musk is often in the spotlight, it's important to remember that Eberhard, Tarpenning and many others also played key roles in founding and building Tesla.

The shareholders and partners

The evolution of Tesla's shareholders has changed massively over the years, particularly through various rounds of financing, partnerships and the company's IPO.

In the first few years after Tesla was founded in 2003, various investors joined the company. Elon Musk became the largest single investor and chairman of the board.

In 2009, Daimler AG acquired around 10% of Tesla's shares, but later sold them.

Toyota also acquired a stake in Tesla in 2010 and collaborated with the company on electric vehicle projects. However, Toyota also sold its shares again in the following years.

Tesla then went public in June 2010, raising around $226 million. The IPO opened Tesla up to a broader group of shareholders, which fundamentally changed the structure of the company. To raise capital for expanding production and other investments, Tesla issued more and more shares in the years that followed.

Over time, large institutional investors such as Vanguard, BlackRock and others also emerged as shareholders in Tesla. Tesla's rise to become one of the most valuable companies in the world also made it part of large indexes such as the S&P 500, resulting in many index funds buying Tesla shares and a large number of individual investors also holding shares in the company.

Why Tesla?

The namesake, Nikola Tesla, born in 1856 in Smiljan, Croatia (then part of the Habsburg Empire) and died in 1943 in New York City, USA, was a Serbian-American inventor and electrical engineer. He is one of the most outstanding minds in the history of electrical engineering and laid the foundations for many modern technologies.

Tesla's most significant contribution to electrical engineering was the development and promotion of the alternating current system, which contrasted with Thomas Edison's direct current system. Alternating current has the advantage that it can be transmitted over long distances and eventually became the dominant power distribution system worldwide.

He developed the Tesla transformer (also known as the Tesla coil) and did important basic research in radio technology. Independently of Wilhelm Conrad Röntgen, Tesla experimented with X-rays and took some of the first X-ray pictures in the USA.

Tesla ended up holding over 300 patents in numerous countries. Although many of Tesla's ideas were considered controversial or too futuristic in his time, they laid the foundation for many modern technologies. He was duly memorialized with the name "Tesla." His legacy lives on in Tesla's mission to advance the transition to sustainable energy.

The long way

The initial phase of Tesla

To understand how Tesla became what it is today, it's helpful to take a closer look at the company's early days.

From the beginning, Tesla's vision was clear: to accelerate the transition to sustainable energy. That meant not only developing electric vehicles, but also renewable energy solutions. And, of course, showing the established once.

The first car: Made in UK

Tesla's first product, four years after its founding, was the Tesla Roadster, introduced in 2008. It was a sporty electric car based on the chassis of the British Lotus Elise and built in England. The Roadster was the first production vehicle to use lithium-ion cells for its battery and offered a range of more than 240 miles per charge. With its sporty design and rapid acceleration (0-60 mph in less than 4 seconds), the Roadster proved that electric vehicles could be not only practical and ecological, but also powerful and attractive.

Nevertheless, the early years were not easy for Tesla. Production of the Roadster was more expensive than expected and there were many technical challenges. The company faced financial collapse several times. Elon Musk, who had already invested significant sums in the company, put more of his own money into Tesla in 2008 just to keep the business running.

Despite these early challenges, interest in Tesla and its products grew. The company soon began developing the Model S, a luxury sedan with an electric powertrain that would hit the market in 2012 and impress both critics and customers with its range, performance and technology.

Tesla's early days were marked by vision, innovation, but also significant challenges. Despite initial setbacks and skeptics, the company became a leading player in the automotive and energy industries. By focusing on electric mobility and renewable energy, Tesla paved the way for broader market adoption of electric vehicles.

Tesla was one of the first companies to focus on developing electric cars for the mass market. This brought with it a host of technical challenges, from battery technology to drive systems.

And with the transition from a niche sports car (the Roadster) to luxury car production (Model S), the company encountered production hurdles and quality issues that were significant.

In addition, the traditional automotive industry and many market analysts were skeptical of Tesla's ability to establish itself as a major new player, especially given the strong competition and the challenges of building a new, previously largely unknown brand.

To make matters worse, there were legal battles in the early years, including a largely obsolete dispute between the co-founders and Elon Musk over who could be called a "founder."

One of the biggest obstacles for electric vehicles (EVs) has always been battery technology. Tesla needed to ensure that their batteries were safe, durable, powerful, and cost-effective. In particular, battery range, charging time, and battery life were critical factors. And as a new automaker, Tesla had to learn how to produce in large quantities, especially with the launch of the Model S and later the Model 3, which saw multiple delays and production bottlenecks.

There continued to be reports of quality issues with some of the early models, ranging from cosmetic issues to functional problems.

To ensure the success of EVs, Tesla also had to invest in a charging infrastructure, which led to the development of the Supercharger network. Building and expanding this network was an immense logistical and financial challenge for the then rather small company, at least compared to the other giants in the automotive industry.

Electric cars have traditionally been more expensive than their gasoline counterparts, mainly because of the high cost of batteries. Tesla therefore had to find ways to reduce costs in order to make their cars accessible to a wider market.

Tesla also chose to incorporate advanced technologies like Autopilot and other software upgrades into their cars. This required constant research and development, as well as navigating regulatory hurdles; another significant problem for the newcomer

As a new automaker, Tesla also struggled to establish stable supply chains, leading to delays in production and delivery.

In the early stages, Tesla had to build the public's image and trust in electric cars. This was a significant challenge in a market dominated by traditional gasoline and diesel cars, where electric vehicles tended to be ridiculed.

These challenges have shaped the company and helped it refine technology and production processes to make it the world-leading electric car manufacturer it is today.

Financial problems, risk of bankruptcy

Tesla faced significant financial difficulties in its early years.

Building a new automotive company requires significant investment in research and development to create a competitive product. Even the development of the first model, the Tesla Roadster, was expensive and complex.

Tesla had difficulties optimizing the production process and producing the Roadster in the planned quantities and at the planned costs. This led to delays and increased costs.

As a startup, Tesla constantly needed new capital to implement its business plans. During the financial crisis of 2008/2009, it became particularly difficult to raise new capital because the financial markets had very limited faith in the chaotic-looking newcomer. Tesla entered a market dominated by established car manufacturers. These companies had the advantage of brand recognition, distribution networks and production capacity.

In Tesla's early years, there was general skepticism about the profitability and practicality of electric vehicles. This affected customer demand and investment.

The cost of batteries, which make up a significant portion of the total cost of an electric car, was also high in the early years. Although prices have come down over time, the initial costs were a major challenge for Tesla.

Tesla naturally had negative cash flows in the early years of its existence as it made significant investments in R&D, production, and sales while revenues were still limited. CEO Elon Musk once mentioned that Tesla was very close to insolvency in 2008. In fact, Musk had to invest much of his own wealth to keep the company afloat.

It was ultimately a combination of innovative products, a vision for the future of mobility and clever financial navigation that enabled Tesla to overcome these early challenges and become what it is today. And, of course, not forgetting the fact that in Elon Musk there was a player ready to invest large sums of money.

Market skepticism

in the first few years after its founding in 2003, Tesla encountered considerable skepticism in both the automotive sector and the financial market. There were several reasons for this skepticism:

Before Tesla, there were quite a few attempts to bring electric vehicles to market. However, some of these earlier models, such as the GM EV1, were discontinued before mass production, leading to skepticism about the long-term viability of EVs.

A major concern of critics was the lack of charging infrastructure. People were concerned about the range of electric vehicles and where and how they would be able to charge their vehicles, especially on longer trips. The cost of batteries was high, and there were concerns about their longevity and performance.

Competition from established manufacturers

Many believed that if large incumbent automakers decided to enter the EV market, they could easily outperform Tesla because they had the necessary resources, vehicle manufacturing experience and market power.

Tesla experienced several financial bottlenecks in its early years. Many market observers were skeptical about whether the company could survive financially.

Despite these challenges, Tesla has slowly but surely changed public opinion through innovative technology, marketing and the charisma of CEO Elon Musk. The Roadster, Tesla's first car, helped establish the image of electric vehicles as powerful and attractive. Later models like the Model S, Model X, and especially the Model 3 expanded Tesla's range and made electric vehicles more accessible to a wider audience.

Over time, and with growing awareness of environmental issues, public attitudes toward electric vehicles have improved significantly. Tesla played a key role in this change by showing that electric vehicles can be both practical and desirable.

Legal disputes among themselves

Tesla, Inc, formerly known as Tesla Motors, was officially founded by Elon Musk, JB Straubel, Ian Wright, Marc Tarpenning and Martin Eberhard. But the road to the company's founding and the years that followed were not without internal tensions and legal battles.

Martin Eberhard was one of the co-founders of Tesla Motors and served as the company's CEO for several years. The relationship between Eberhard and Musk, who was one of the company's main investors in its early days, deteriorated over time.

Eberhard was removed from his position as CEO in 2007 and eventually left the company. In 2009, Eberhard filed a lawsuit against Musk and Tesla, claiming Musk attempted to "override" his contributions to Tesla and forced him out of the company. Eberhard also claimed that he was given false information about the company's financial health and that he was unlawfully removed from his position.

Among the most contentious issues was how the "founder" designation was used. Eberhard claimed that he and Tarpenning were the true founders of Tesla, while Musk came along later. Musk argued that he was instrumental from the beginning, both as an investor and as a strategic decision maker.

The lawsuit was eventually settled out of court. One result of the settlement was that both Eberhard and Musk were recognized as "co-founders" of Tesla.

Google as savior?

There were reports that Google almost acquired Tesla in 2013. At the time, Tesla was in financial trouble and Elon Musk was in talks with Google's Larry Page about a possible takeover.

The terms of the deal would have called for Google to pay $6 billion for the acquisition and provide an additional $5 billion for Tesla's expansion. Musk would have retained control of the company

Ultimately, however, the deal did not materialize, as Tesla's sales figures for the Model S skyrocketed within a short time and the company achieved a turnaround. This made the need for a takeover obsolete.

What if...?

The acquisition of Tesla by Google is a hypothetical scenario that could have had many implications for both companies and possibly the entire technology and automotive industries. As this is purely speculative, it is difficult to make accurate predictions, but we can discuss some potential outcomes:

The technologies of both companies could be uniquely combined. Google already has strong data analytics, AI and software development capabilities, while Tesla has been a leader in electric mobility and autonomous driving technologies. A merger could lead to accelerated innovation in these areas, such as integrating Google Assistant into Tesla cars or using Google's data infrastructure to improve Tesla's autonomous driving systems.

The combination of the two companies' market power could lead to an almost insurmountable position in their respective markets. Tesla could benefit from Google's extensive resources and global presence to expand faster and outpace competitors. Google, in turn, could solidify its presence in the emerging market for autonomous vehicles and electric mobility.

Such a merger would likely face significant regulatory hurdles, particularly with regard to antitrust and competition law. Regulators may have concerns about the potential monopoly and market power that the merger could create.

Although both companies are considered innovative and disruptive, they have very different corporate cultures and leadership styles. Elon Musk is known for his risky but visionary approach, while Google, under the Alphabet umbrella, has a more conservative and structured corporate culture. These differences could lead to tensions and implementation difficulties.

Shareholders, employees and customers of both companies could be affected by the acquisition in a variety of ways. While some stakeholders may benefit from increased financial stability and synergies, others may have concerns about data privacy, job security or product quality.

A potential merger could also have far-reaching implications for issues such as sustainable mobility and technology ethics. Tesla and Google could pool their resources to make progress in areas such as renewable energy or sustainable manufacturing, but there could also be concerns about the impact of such a large corporation on society.

Elon Musk

Elon Musk: Visionary of the 21st century

Elon Musk, born in Pretoria, South Africa, in 1971, is undoubtedly one of the most outstanding personalities of the 21st century. With a series of innovative companies, he has initiated groundbreaking changes in various industries.

Under Musk's leadership, Tesla has evolved from a start-up to one of the world's leading electric car manufacturers. With models like the Model S, Model 3, Model X and Model Y, Tesla has revolutionized the automotive industry and accelerated the transition to sustainable mobility.

Space Exploration Technologies Corp, better known as SpaceX, aims to colonize space. The company has focused on reusability of rockets and spacecraft with its

Falcon and Dragon series, which should make space travel more cost-effective.

The Boring Company: With this company, Musk wants to tackle the traffic problem in big cities by building underground tunnel systems.

Neuralink: This biotech company focuses on developing brain-computer interfaces that aim to connect human brains to computers.

SolarCity: Although acquired by Tesla in 2016, SolarCity began as an independent company that changed the market for solar energy and residential solar storage.

Musk is known not only for his entrepreneurial activities, but also for his controversial and often provocative statements on platforms such as Twitter (now "X"). He has established himself as a critic of traditional ways of thinking and a promoter of innovative approaches. His work ethic, vision and willingness to take risks have made him one of the most influential people of our time. Despite criticism and headwinds, Elon Musk remains a driving force behind some of the most significant technological breakthroughs of our generation.

Money and partners

Participations and cooperations

Tesla has collaborated with a number of companies in various industries over the years. One of the best-known examples is the partnership with Panasonic in the field of battery technology. Together, they built the Gigafactory in Nevada, where batteries for Tesla vehicles and energy storage products are manufactured.

Daimler AG, the parent company of Mercedes-Benz, invested in Tesla in 2009 and worked on the development of the electric powertrain for the Mercedes-Benz B-Class Electric. However, Daimler later sold its Tesla shares again.

Toyota invested about $50 million in Tesla in 2010 and formed a partnership to develop electric vehicles. Tesla supplied the electric powertrain for the Toyota RAV4 EV. However, the partnership ended and both companies sold their respective shares to each other.

Although SolarCity is a company founded by Elon Musk's cousins and later acquired by Tesla, it represents an important collaboration in the renewable energy sector.

Tesla has also worked with NVIDIA to develop powerful, AI-driven computers for its Autopilot and Full Self-Driving features.

China's Contemporary Amperex Technology Co, Limited (CATL) is another key partner for Tesla in battery technology, especially in the emerging Chinese market.

Deutsche Bank and other financial institutions - Tesla also has collaborations with financial institutions for various financing programs, including leasing and loan options for customers.

Tesla is also working with a number of software companies, including Google for maps and navigation services, to improve the user experience in its vehicles.

Tesla also has partnerships with utilities to distribute and deploy its energy storage products such as the Powerwall and Powerpack.

Mercedes-Benz

The partnership between Tesla and Mercedes-Benz represents a significant moment in the history of both companies. The highly publicized cooperation began in 2009, when Daimler made an investment of about $50 million in Tesla, which was a lagging company at the time. At the time, Tesla was still a relatively young and small company that had just begun to realize its vision for electric cars. Daimler, on the other hand, was a decidedly established automaker with a long history and deep pockets. The investment gave Tesla not only much-needed capital, but also some legitimacy in the automotive industry.

The main focus of the partnership was on the development of electric powertrains and the electrification of existing Mercedes-Benz models. One concrete result of this cooperation was the Mercedes-Benz B-Class Electric, a fully electric vehicle that was launched in 2014. Tesla was responsible for supplying the electric powertrain and battery, while Mercedes-Benz handled the design, manufacturing and distribution of the car. Although the B-Class Electric did not become as popular as other models, it served as a testing ground for both companies to gain experience in the development and production of electric cars.

In addition, the partnership had symbolic significance. It showed that traditional automakers were taking electromobility seriously and were willing to work with startups in this field. For Tesla, the cooperation also brought a number of benefits. In addition to the financial investment, Tesla gained access to Daimler's extensive manufacturing expertise, which was invaluable to the young company. Furthermore, the partnership helped improve Tesla's brand image by establishing the company as a serious player in the automotive industry.

However, the partnership did not last. Daimler sold its Tesla shares in 2014 for about $780 million, which represented a significant gain on the original investment. Both companies began to forge their own paths in electric mobility, with Tesla focusing on a wide range of electric vehicles, from luxury cars to more affordable models, and Mercedes-Benz developing its own electric car strategy, which led to the launch of the EQ range and other models.

Did Mercedes' exit make sense?

Whether or not the exit of Mercedes-Benz (or Daimler AG) from its partnership and investment in Tesla was sensible is a question that can be viewed from different perspectives. Here, too, the question arises: What if...?

Daimler sold its Tesla shares in 2014 for around $780 million, having originally invested around $50 million. From a purely financial perspective, Daimler made a considerable profit on the sale. However, if one takes into account that the value of Tesla has risen sharply in the following years, it could be argued that Daimler could have profited even more if it had kept its shares.

During the partnership, Mercedes had access to Tesla's electric drive technology, but after the exit, Mercedes began to invest more in its own research and development. This can be seen as positive, as it allowed Mercedes to develop an independent technology platform that is better aligned with its own market position and strategy.

Both companies have very strong but different brands. Tesla is known for its innovation and disruption, while Mercedes-Benz stands for its long tradition of quality and luxury. A continued partnership might have diluted the brand image of both companies.

In retrospect, it can be argued that the exit was both sensible and of limited success, depending on which factors one considers important. From a financial perspective, Daimler certainly benefited, but may have benefited more. Strategically and technologically, the exit allowed both companies to go their own ways, which can be seen as beneficial given the rapid developments in the field of electromobility. On the other hand, it must be noted that Tesla is now one of Mercedes' most important competitors. However, it is doubtful whether this development of Tesla could have been prevented by a minority shareholding alone.

Toyota

The partnership between Tesla and Toyota began in 2010, when Toyota, a world-leading automaker with a long history of developing hybrid technology, invested about $50 million in Tesla. This investment was not only an important financial injection for Tesla, but also a vote of confidence from one of the world's largest and most respected automakers.

The main project resulting from this partnership was the development of the electric Toyota RAV4 EV. Tesla was responsible for supplying the electric powertrain and battery technology, while Toyota handled the vehicle design and manufacturing.

The RAV4 EV was launched in 2012 and was intended to play an important role in demonstrating the feasibility and acceptance of electric vehicles in the mass market.

The partnership offered both companies important synergies. Tesla benefited from Toyota's many years of experience in large-scale vehicle production, while Toyota was able to benefit from Tesla's innovative electric drive technology. This enabled Tesla to refine its technical know-how while stabilizing its financial position. Toyota was able to accelerate its own electric mobility research through the collaboration, while also having the opportunity to better understand the electric vehicle market.

Despite the initial enthusiasm and joint projects, both companies eventually sold their shares to each other and went their separate ways. Tesla, meanwhile, had accumulated enough capital and expertise to develop a wide range of electric vehicles on its own. Toyota, meanwhile, continued its own path, continuing to invest in various forms of electric mobility, including fuel cell and hybrid technologies. Toyota focused more on hybrid powertrains as well, while Tesla famously focused on all-electric powertrains.

The partnership was a success in many respects. Even if the RAV4 EV did not become a bestseller, it served as a valuable field of experimentation for both companies.

Panasonic

The collaboration between Tesla and Panasonic represents one of the most distinctive and long-lasting partnerships in the electric car industry. It officially began in 2010, when the two companies reached an agreement to jointly develop lithium-ion battery cells for electric vehicles and continues to this day. This partnership came at a critical and highly important time for Tesla, when the company was working on the development of the Model S, its first mass-market electric car.

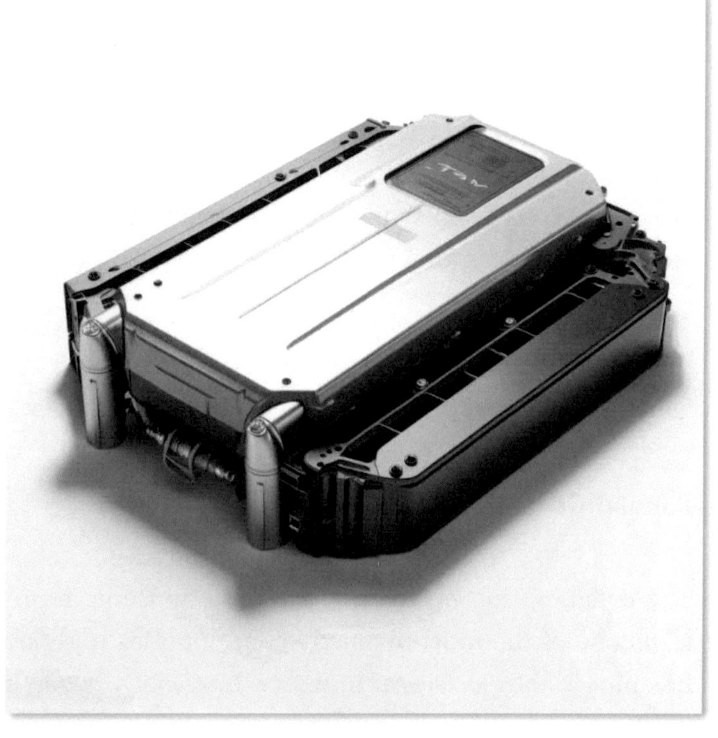

The most ambitious milestone of the partnership was undoubtedly the construction of the Tesla Gigafactory in Nevada, USA. This huge facility, the construction of which began in 2014, was developed to enable the mass production of novel battery cells and packs. Panasonic was a key partner in this, taking on the role of main supplier and manufacturer for the battery cells, while Tesla took on the architecture of the battery modules and packs. The Gigafactory was intended not only to reduce production costs, but also to increase security of supply by minimizing dependence on external suppliers.

Panasonic, a company with many years of experience in battery technology, benefited from Tesla's innovative strength and market access. Tesla, in turn, benefited from Panasonic's expertise in battery manufacturing and technology. These synergies led to significant advances, for example in the batteries' energy density, durability and safety.

Despite the strong partnership, both companies have also had to make adjustments over the years. Panasonic initially struggled to keep up with Tesla's high demand and high quality standards, which caused some tension. Tesla also began exploring other battery cell suppliers, including companies like LG Chem and CATL, to diversify its supply chain and minimize risk.

Panasonic is still an important partner for Tesla, especially for battery cell production at the Gigafactory in Nevada. However, there are also indications that Tesla is pursuing plans to develop and produce its own battery technology, which underscores the dynamic nature of this partnership.

Acquisitions by Tesla

Tesla has consistently acquired other companies to expand its technology, manufacturing capabilities and expertise in a variety of areas. Some of the most notable acquisitions are:

SolarCity

In 2016, Tesla acquired solar company SolarCity for about $2.6 billion. This was one of Tesla's largest and most controversial acquisitions, but it was closely linked to Elon Musk's vision of creating an integrated energy company. Through this acquisition, Tesla wanted to conquer the renewable energy market and offer not only electric cars, but also solar energy solutions.

Maxwell Technologies

In 2019, Tesla acquired Maxwell Technologies, a manufacturer of ultracapacitors and batteries, for about $218 million. The acquisition was seen as a strategic move to enhance Tesla's battery technology capabilities. Maxwell has a patented dry electrode manufacturing technology that could boost the performance of lithium-ion batteries.

Perbix

Tesla acquired Perbix, a company specializing in automation, in 2017, and although financial details were not made public, the acquisition served to improve Tesla's production capabilities, particularly in terms of manufacturing facilities for its electric cars.

Hibar Systems

In 2019, Tesla acquired Hibar Systems, a Canadian company specializing in battery cell production systems. This acquisition fit well with Tesla's efforts to increase its own battery production and reduce its dependence on suppliers.

DeepScale

Also in 2019, Tesla bought DeepScale, an artificial intelligence startup that specializes in optimizing deep learning models for use in autonomous vehicles. DeepScale's technology could help improve the capabilities of Tesla's Autopilot system.

SilLion

Although less well known, Tesla also acquired SilLion, a company with expertise in battery materials, in 2016. This acquisition is believed to have supported Tesla's efforts in battery technology.

The strategy behind these acquisitions is clear: Tesla wants to control the entire value chain, from energy generation to battery production to car production and autonomous driving capabilities. Each of these acquisitions plays a key role in Tesla's overall strategy to integrate the various aspects of sustainable energy and mobility.

From a German perspective, the recent acquisitions of Grohmann, Wiferion and ATW are still significant.

Tesla and the stock market

Tesla is one of the most eye-catching and talked-about companies on the stock market, and its share price has experienced remarkable volatility and growth momentum over the past decade. The company went public in 2010 and quickly became one of the most valuable automakers in the world, despite comparatively low production numbers compared to established automotive giants like General Motors, Toyota, and Volkswagen.

One of the main reasons for Tesla's stunning rise on the stock market is the vision and charisma of its CEO Elon Musk, who has positioned the company as more than just a manufacturer of electric vehicles. Tesla is seen as a technology company that wants to drive the transition to a sustainable energy future. This includes not only electric cars, but also energy storage solutions and solar products. This broad vision has attracted investors who believe in a long-term transformation of the energy and transportation landscape.

However, the company is not without controversy and has a very volatile stock market history. Elon Musk himself has made headlines several times, for example with tweets that have affected the share price, or with unconventional business decisions that have often led to sharp fluctuations in the share price. In addition, Tesla's valuations are always up for debate, with critics arguing that the company is overvalued in relation to its actual financial results and production figures.

Despite the skepticism, Tesla has achieved impressive financial results in recent years. The company became profitable and was able to significantly increase its production, especially after the start-up of the gigafactories in Nevada and Shanghai. Tesla has also invested in research and development of battery technologies, which is seen as another catalyst for growth.

Tesla's high valuation and success on the stock market also have implications for the entire automotive industry. It has increased the pressure on established manufacturers to invest in electric vehicles and sustainable technologies. Some investors and market observers even view Tesla as a benchmark for success in the emerging electric vehicle market.

Overall, Tesla is a phenomenon on the stock market that attracts a lot of attention and has both supporters and critics. It symbolizes the rapid changes in the automotive industry and the broader transition to a more sustainable global economy. With its focus on innovation and disruption, Tesla has certainly set new standards in the investment world, and it remains exciting to see how the company will continue to evolve in the coming years.

Tesla and the money

Tesla has had a remarkable financial journey, moving from initially high losses to sustained profitability. As a start-up company in the capital-intensive automotive industry, Tesla initially struggled with significant costs for research and development, production and market launch of its electric vehicles. As a result, the company was not profitable in its early years.

The road to profitability has been challenging for Tesla, requiring significant investments in product development, production capacity and infrastructure. A turning point came in the third quarter of 2018, when Tesla reported its first ongoing profit in a long time. This was due in part to the successful production launch of the Model 3, Tesla's first mass-market electric vehicle. Sales and production numbers rose steadily in subsequent quarters, and Tesla maintained profitability in several quarters.

Tesla was then able to achieve sustained profitability from 2020 onwards. This year, despite the challenges of the COVID 19 pandemic, the company managed to be profitable in all quarters. This continued financial success was the result of a combination of factors: increased production efficiencies, successful new model launches such as the Model Y, an increase in sales particularly in new markets such as China, and revenue from the sale of emission rights to other automakers.

In addition to vehicle sales, Tesla has also developed other revenue streams, including energy storage products and solar through its acquisition of SolarCity. In addition, Tesla offers software upgrades for its vehicles, including the "Full Self-Driving" (FSD) package, which generates additional high-margin revenue.

Despite or because of its profitability, Tesla remains a company that is investing heavily in its future. It has invested billions in building new production facilities, including the Gigafactory in Nevada, the Gigafactory in

Shanghai, and future sites in Berlin and Texas. These investments could weigh on profitability in the short term but are critical to the company's long-term growth plans.

In summary, after years of losses and intensive investments, Tesla has now reached a phase of sustainable profitability. The company's financial situation has stabilized, and it has shown that it can sustainably generate profits, which boosts investor confidence and further drives the company's share valuation.

Could Tesla have been founded in Europe?

The question of whether Tesla could have been successfully established in Europe is complex and depends on a number of factors, including the investment climate, regulatory environment, talent pool, and market access. Below are some thoughts on this hypothetical scenario:

Silicon Valley in the U.S. offers a unique ecosystem of venture capital that provides significant financial resources to companies like Tesla in their early stages. European countries also have growing venture capital markets, but these are often less aggressive and risk-averse than their U.S. counterparts. This may have hampered the rapid rise of a disruptive company like Tesla.

Regulatory hurdles for the automotive industry in Europe are high, especially in terms of environmental regulations and safety standards. While these regulations could be beneficial for a company promoting sustainable mobility, they could also slow down the speed of innovation and market entry.

Another advantage of Silicon Valley is access to a highly skilled talent pool in areas such as software development, engineering, and entrepreneurship. While Europe also has world-class universities and research institutions, the market is more fragmented and access to talent is more difficult.

The U.S. market offers a large, relatively homogeneous consumer base that allows companies to scale quickly. Europe consists of a variety of markets with different languages, cultures, legal systems, and consumer habits, which could make expansion more complicated.

The entrepreneurial culture in the U.S. tends to be more risk-averse and tolerant of failure, which is an important factor for startups. European countries have made progress in this regard in recent years, but the culture of entrepreneurship in Europe has traditionally been less aggressive.

While it would theoretically be possible for a company like Tesla to be established in Europe, the challenges and conditions would be different and could affect the speed and scale of growth. It would have required significant adjustments in business strategy, financing, and market entry to be as successful in Europe as in the US.

Tesla International

The international networking of the automotive industry is a complex web of supply chains, joint ventures, partnerships and agreements that span continents. This globalization has enabled manufacturers to benefit from economies of scale, gain access to new markets and implement innovations more quickly.

Supply chains are a prime example of this interconnectedness. A single automobile can consist of thousands of components that come from different parts of the world. For example, the electronics could come from Asia, the engines from Germany and the body parts from the USA. This international distribution of production allows manufacturers to cut costs while benefiting from specialized expertise.

In addition, strategic partnerships, and alliances between car manufacturers from different countries are common. For example, European, American, and Asian companies often work together in areas such as electromobility, autonomous driving or connected car technologies. These collaborations not only accelerate the development of new technologies, but also help to share the enormous research and development costs.

However, international networking also has its challenges, particularly regarding regulatory differences between different markets. Emission standards, safety requirements and customs duties can vary from country to country, increasing complexity for globally operating automakers.

Overall, the international interconnectedness of the auto industry has led to unprecedented efficiency and innovation, but it has also increased the complexity of the industry and made it vulnerable to global risks, such as trade wars or supply chain disruptions. It remains a key factor in the ongoing transformation and evolution of the global automotive industry.

Tesla sells its cars in many countries around the world, focusing on key markets such as the USA, Europe, and China. In Europe, countries such as Norway, Germany and the Netherlands are particularly important sales markets. In Asia, China is by far the largest and most important market for Tesla. The company even has its own production facility, the Gigafactory Shanghai, to better serve the Chinese market.

In addition to electric vehicles, Tesla also offers other energy solutions such as solar panels and energy storage. These products are also sold internationally, but to a lesser extent than the vehicles.

Tesla operates several production facilities around the world. In addition to plants in the U.S. (Fremont, Nevada) and China (Shanghai) and Germany (Brandenburg), Tesla has announced plans for additional Gigafactories. These international manufacturing locations will allow Tesla to serve local markets more efficiently and minimize import tariffs.

Another aspect of international positioning is adaptation to local legislation and standards. Tesla, for example, has developed special versions of its cars for the European or Chinese markets to comply with the regulations in force there.

Tesla has an extensive sales and service network that covers the major markets. There are showrooms and service centers on every continent, even in countries where electric mobility is still in its infancy.

Tesla in Europe

Tesla's presence in Europe has expanded significantly in recent years, and the continent has become an important market for the company. The European market has some attractive characteristics that make it particularly relevant for Tesla:

Europe has a growing demand for electric vehicles, driven by government incentives, stringent emissions regulations, and increasing consumer environmental awareness. Countries such as Norway, the Netherlands and Germany are leaders in electric mobility and important markets for Tesla.

The regulatory environment in Europe is conducive to electric cars. Many European countries have ambitious plans to restrict or completely ban the sale of gasoline and diesel vehicles in the coming years. This creates a favorable climate for companies like Tesla that focus on sustainable mobility.

Tesla is building a Gigafactory near Berlin, Germany, to increase its production capacity in Europe. This production facility should enable the company to better respond to local demand, reduce logistics costs and meet specific needs of European customers.

Tesla has established sales and service centers in key European cities. The Supercharger infrastructure, which Tesla has expanded throughout Europe, facilitates long-distance travel and is an important factor in the acceptance of electric vehicles.

While Tesla has a dominant position in the U.S. in terms of electric cars, competition in Europe is more intense, with strong players such as Volkswagen, BMW and Mercedes-Benz entering or having already entered the electric vehicle market. This competition could affect Tesla's growth prospects in Europe, but also provides a spur to innovation and improvement.

In summary, Europe is a core market for Tesla that offers significant growth potential. The combination of high environmental standards, government incentives and growing consumer demand for electric vehicles makes the continent a strategically important market for the company. Tesla is therefore investing heavily in expanding its presence in Europe, both in terms of production and sales and service.

Tesla in Asia

Tesla's presence in Asia has increased significantly in recent years, with the focus of its activities mainly on the Chinese market.

China is a key market for Tesla and for the electric car industry as a whole. It is the largest automotive market in the world and has set aggressive goals to promote electric mobility. Tesla has recognized this and opened a Gigafactory in Shanghai in 2019. This factory has several advantages: it allows Tesla to serve the Chinese market faster, lowers production costs, and reduces the impact of trade barriers or tariffs. The success of the Model 3 and Model Y vehicles in China has helped establish Tesla as a leader in the electric vehicle segment in the country.

Tesla has also expanded its presence in Japan and South Korea, albeit to a lesser extent than in China. These countries have highly developed automotive industries and growing acceptance of electric vehicles, especially in urban areas. Tesla sees these markets as opportunities for growth and has opened sales and service centers there.

Tesla's presence in India is still in the development stage, but the company has plans to gain a foothold in this potentially huge market. Given air pollution problems and the government's initiative to promote electric vehicles, India could become an interesting market for Tesla. However, low average incomes and infrastructure pose challenges.

In addition to these core markets, Tesla also has operations in other Asian countries such as Singapore, Hong Kong and the Philippines. In these smaller markets, Tesla focuses primarily on selling vehicles and building a charging infrastructure.

Tesla's strategy in Asia is clearly focused on growth and market share gains, with China as the most important market in the region. The importance of China to Tesla is such that the company's performance in this market has significant implications for its global health. Tesla's operations in other Asian countries are less pronounced, but the company views the entire region as critical to its long-term ambitions and is investing to expand its presence and infrastructure.

Tesla in South America

Tesla's presence in South America is much more limited compared to other regions such as North America, Europe, or Asia. There are a number of factors influencing Tesla's entry and growth in South American markets:

South America is a heterogeneous region with a mix of development levels, economic climate, and infrastructure. In general, average incomes are lower than in Tesla's core markets, which could potentially limit sales of premium electric vehicles.

Some countries in South America have introduced incentives for renewable energy and electric mobility, but these measures are often less comprehensive than in Europe or Asia. This could limit demand for electric vehicles and thus restrict Tesla's growth in the region.

Electric car infrastructure, especially in terms of charging facilities, is less developed in South America. This could be a challenge for Tesla, but it could also mean an opportunity to launch their own Supercharger networks.

Tesla has not yet established a major official sales or service channel in South America. However, in countries such as Brazil, Argentina and Chile, there are a growing number of Teslas entering the country through independent importers. Owners of these vehicles often face maintenance and service challenges due to the lack of official Tesla service centers.

While South America is not currently considered a core market for Tesla, the region could become more important in the future. Countries such as Brazil and Argentina have large amounts of natural resources such as lithium, which is used in electric car batteries. Strategic partnerships or investments in these countries could both strengthen Tesla's supply chain and boost the local electric vehicle market.

The vehicles

Tesla Roadster (2008)

The 2008 Tesla Roadster marks a significant milestone in the history of the automotive industry. It was Tesla's first production vehicle and served as a poster child for the potential of electric mobility. At a time when electric vehicles were often considered impractical or an unattractive alternative to conventional gasoline-powered cars, the Roadster broke those stereotypes and set new standards. It was proof that Tesla could build cars.

Concept

The basic concept behind the 2008 Tesla Roadster was clear and simple: to convince the world that electric vehicles can be not only practical and environmentally friendly, but also powerful and stylish. Tesla's strategy was to create a niche product for the luxury market that also demonstrated the technological possibilities of electric vehicles. The Roadster was promoted not only as an environmentally friendly vehicle, but also as a high-performance sports car capable of competing with established brands in the market.

The design of the 2008 Tesla Roadster was inspired by classic sports cars, but also offered a modern and fresh look. It was unmistakably based on the Lotus Elise platform, with Tesla making many modifications to accommodate the electric drive and battery. The roadster's lines were sleek and flowing, giving it a timeless look. The convertible design emphasized the car's sporty character.

In terms of interior space, the design of the Tesla Roadster was functional rather than opulent, with a focus on the driving experience. The interior was cramped and driver-focused, with all controls easily accessible and intuitively located. Instrumentation and battery monitoring were still very rudimentary and today are more reminiscent of an e-bike than a modern EV.

Still, the 2008 Tesla Roadster was far more than just another electric car on the market. It was a bold statement by Tesla and Elon Musk about what was possible in the world of electric cars. By combining a powerful electric powertrain with a stunning design, the Roadster proved that electric cars are not only practical, but desirable. It was the beginning of Tesla's journey and laid the foundation for what the company would achieve in the years that followed.

The 2008 Tesla Roadster was based on electric drive technology and used lithium-ion batteries to store electrical energy for the drive motor.

Tesla used the Lotus Elise as the basis for the Roadster but modified it significantly to accommodate the electric components. The Roadster featured a 3-phase 4-pole electric motor with a maximum torque of 400 Nm. This motor enabled the Roadster to accelerate from 0 to 60 miles per hour (0 to 97 km/h) in less than 4 seconds.

The Roadster's battery pack consisted of about 6,831 individual lithium-ion cells, similar to those used in laptop computers. This battery pack provided the Roadster with a range of about 244 miles (393 km) according to the EPA cycle at the time.

The Roadster used a power electronics module (PEM) to manage the energy between the battery pack and the engine. This module also controlled the charging of the batteries.

Like many electric vehicles, the Roadster had a regenerative braking system that fed energy back into the batteries during braking.

With the included home charging kit, the Roadster could be fully charged in about 3.5 hours.

Aside from its impressive acceleration, the Roadster also had a top speed of 125 mph (201 km/h).

The Roadster was not intended for the mass market. It was produced in limited numbers and sold at a high net price, ranging from $100,000 to $130,000 depending on equipment and region.

Tesla managed to attract considerable media attention, both because of the innovations of the Roadster and because of its charismatic co-founder and CEO Elon Musk. This helped the company position itself as a premium brand in the electric vehicle sector.

Despite the initial success and media attention, Tesla encountered significant challenges in the production and delivery of the Roadster. There were delays, cost overruns and quality issues that the company had to address.

The Roadster was just the first step in Tesla's long-term plan. It was intended to generate money and interest in electric vehicles, which it undoubtedly did. Tesla used the revenue and lessons learned to invest in larger and more affordable models like the Model S and Model X.

The Tesla Roadster was built in collaboration with Lotus at their plant in Hethel, England. Tesla supplied the electric drive components, while Lotus contributed most of the vehicle design and manufacturing, based on their own sports car, the Lotus Elise. This gave Tesla the opportunity to benefit from Lotus' manufacturing expertise and reduce development costs.

In terms of units, the Tesla Roadster was produced from 2008 to 2012, with a total of about 2,450 units manufactured.

The Tesla Roadster was received overwhelmingly positively, both by the press and by customers. The launch of this vehicle marked a significant turning point in the public perception of electric vehicles.

The introduction of the Roadster came at a time of growing awareness of environmental issues and, in particular, the impact of greenhouse gas emissions on climate change. Many saw the Roadster as an environmentally friendly alternative to conventional gasoline-powered sports cars.

The vehicle attracted the attention of many media outlets, both in the automotive industry and in general news sources. The interest in Elon Musk as an entrepreneur and the fact that this was Tesla's first car increased media coverage.

Despite the largely positive reception, there was also skepticism and criticism. Some critics pointed out that the Roadster's high price made it inaccessible to the average buyer. Others expressed concerns about the long-term reliability of electric vehicles and charging infrastructure.

Tesla struggled with production delays and initial quality issues noted by some buyers and media. These challenges led to mixed reviews in terms of customer satisfaction and company performance in the early years.

Commercial result

Directly with the Tesla Roadster, the company did not make any money in an overall view, mainly due to high development costs, production challenges and overall low production volumes. The Roadster was Tesla's first model and, in many ways, served as a proof-of-concept for electric vehicles with high performance and range.

As a new company, Tesla had to make significant investments in developing the necessary technology and infrastructure. This included battery technology, the drive system and other components specific to the Roadster.

There were initial production challenges with the Roadster, including quality issues and delays that added costs. In addition, the Roadster was produced in relatively small quantities, which made it difficult to recoup the initial development costs.

More important than direct profits was the Roadster's role as a catalyst for the company. It showed that electric vehicles could be powerful and practical. This helped generate investor interest and confidence, which in turn led to further investment in Tesla and the development of subsequent models, such as the Model S.

Overall, the Roadster was more of a means to an end than a revenue-generating product for Tesla. It played a crucial role in putting the company on the map, attracting investors, developing technology, and paving the way for future mass-market oriented models. Tesla's true financial success came later with models like the Model S, Model X and especially the Model 3.

The roadster as an enthusiast vehicle

The original Tesla Roadster has achieved high cult status among collectors and car enthusiasts. It was built in approximately 2,450 units from 2008 to 2012. Because it was Tesla's first production vehicle and was only produced in limited numbers, it has historical value for many as one of the vehicles that heralded the turn to electric mobility like no other. Today, a well-preserved Roadster is often worth more than its original price at the time - another indication of the great importance of this vehicle, and not just for Tesla.

Tesla Model S

The Tesla Model S is a fully electric luxury car that serves as Tesla Motors' flagship model to this day. Since its launch in 2012, it has paved the way for electric vehicles in the mainstream and has often been considered a benchmark for other electric cars.

Development

The Model S was the first vehicle that Tesla developed entirely in-house. While the Roadster showed that electric vehicles could be powerful and fun to drive, the Model S was meant to prove that an electric car could also be practical, luxurious, and suitable for the mass market. Elon Musk wanted to build a car that could compete with the big players in the industry in every way, not just in terms of electrification. Development took several years and cost hundreds of millions of dollars, but the end result was a groundbreaking car that redefined many industry standards and has been sold essentially unchanged since 2012. That's another indicator of the superior quality of this car.

Design

The design of the Tesla Model S is a seamless blend of aesthetics, functionality and innovation that makes it one of the most distinctive electric vehicles on the market.

The exterior appearance of the Model S is characterized by a simple but elegant design language. The car dispenses with excessive decoration or complicated design elements and instead presents itself as clean and modern. This minimalist approach draws attention to the car's fine lines and curves, creating a timeless appearance. Nevertheless, the recognition value is high.

The slim shape of the Model S is not only aesthetically pleasing, but also functional. The design was intensively tested in the wind tunnel to keep the drag coefficient as low as possible. Low drag is crucial to the efficiency of an electric vehicle, as it increases range and reduces energy consumption.

The minimalist approach continues inside as well. The cockpit is dominated by a large central touchscreen that replaces most of the physical buttons and switches. This simplifies the user interface and enables more intuitive interaction with the vehicle. In addition, the interior offers a surprising amount of space for passengers and luggage thanks to the compact electric drive components and the flat battery pack under the floor.

When it comes to materials, Tesla focuses on quality and sustainability. From the premium seats to the trim, high-quality materials are used that are not only luxurious but also durable. Tesla has been replacing leather materials with synthetic, vegan alternatives since 2017. Incidentally, this applies to all models.

The design of the Model S cleverly integrates technological elements such as cameras, sensors, and radar for the autopilot functions. These elements are inserted into the body in such a way that they are barely noticeable, contributing to the car's sleek and uninterrupted exterior appearance.

The design of the Tesla Model S is a prime example of the fusion of form and function. Every element, from aerodynamics to material selection, serves a purpose and contributes to the overall performance and experience of the vehicle. It's a design that is not only aesthetically pleasing, but also forward-thinking and innovative. In an industry often dominated by traditional norms and values, the Model S is a refreshing departure and stands out as a distinctive masterpiece of modern automotive design. The design is so good that it has been produced largely unchanged for over 10 years now and still looks attractive, striking and yet elegant.

Powertrain and battery technology

The powertrain and battery technology of the Tesla Model S are at the heart of its performance and efficiency. These components are at the heart of the electric vehicle and are critical to its outstanding driving characteristics, range, and low energy consumption. The key features of these technologies are explained in more detail below.

Depending on the model, the drivetrain of the Tesla Model S consists of one or more electric motors that are directly connected to the wheels. Some versions of the Model S even use two motors to enable all-wheel drive. These electric motors are remarkably efficient and can deliver impressive acceleration that rivals high-end sports cars. Direct drive also allows for very precise control of motor output, which improves driving stability and handling. In addition, electric motors require less maintenance than internal combustion engines because they have fewer moving parts.

The battery is perhaps the most talked-about feature of electric vehicles, and the Tesla Model S sets the standard here as well. The vehicle uses lithium-ion cells arranged in a specially designed, low-profile battery pack under the vehicle floor. This configuration has several advantages:

The underfloor battery arrangement lowers the vehicle's center of gravity, which improves roadholding.

Since the battery is located under the floor, the interior of the vehicle is not affected, which creates more space for passengers and luggage.

The battery pack is equipped with an advanced cooling system that regulates the temperature of the cells during charging and discharging to extend battery life.

The batteries in the Model S are designed for long ranges. Newer models can achieve a range of over 600 kilometers or 400 miles, which is more than sufficient for most driving needs.

Tesla's Supercharger technology makes it possible to charge the battery up to 80% in about 30 minutes. The extensive network of Supercharger stations makes it so easy and convenient to cover even longer distances. According to the latest surveys, the batteries maintain 90% performance over a mileage of 300,000 kilometers. Degradation is around 5% during the first 100,000 kilometers and then levels off.

Taken together, the Tesla Model S powertrain and battery technology create a synergy of performance, efficiency and ease of use that was unheard of at the time. They are the result of years of research and development and set the standard for what is possible in the world of electric vehicles. The constant evolution of these key technologies has kept the Model S at the forefront of a rapidly changing industry to this day.

Software and autopilot

One of the most revolutionary features of the Model S is the ability to receive over-the-air updates. These updates can include anything from minor bug fixes to significant software upgrades that unlock new features. Autopilot, Tesla's advanced driver assistance system, is also an important part of the technology ecosystem. With features like lane change assist, adaptive cruise control, and even a "summon" feature that allows the car to autonomously drive to its owner, Autopilot is setting new standards in the industry. More on the cross-model topic later.

Interior and user interface

A central touchscreen serves as the control center for nearly all the car's functions, from climate control to navigation. This interface is intuitively designed and allows the driver to concentrate on the road while he or she has the car's full range of functions at their disposal.

Security technology

The Model S is also cutting-edge in terms of safety. It has a host of sensors and cameras that, combined with advanced software, enable a range of safety features. These include automatic braking, collision warning and a robust design that allows the vehicle to score excellently in various safety tests.

The Model S is more than just a new car; it is a comprehensive, networked mobility system that sets the course for the future of personal transportation and is constantly being equipped with innovations. This is also new territory: A Tesla basically does not become obsolete, as it is constantly being updated and thus adapted to new developments. Only the hardware - which is relatively timeless in electric cars anyway - remains the same.

Reception and influence

Since its launch, the Model S has won numerous awards and has often been considered one of the best electric cars on the market. It has not only changed the public perception of electric vehicles, but also pushed established car manufacturers to develop their own electric vehicles.

Overall, the Tesla Model S has revolutionized the automotive industry in many ways, from the technology and performance an electric car can offer to the way cars are sold, updated, and repaired.

Tesla Model X

The Tesla Model X is a fully electric SUV built on the Model S. From its iconic "falcon wing" doors to its advanced Autopilot technology, the Model X has influenced the automotive market in many ways.

The Model X was introduced in 2015, a few years after the success of the Model S. Tesla wanted to conquer the luxury SUV segment while taking advantage of electrification. It was designed with the goal of creating a practical, spacious, yet powerful vehicle that also minimizes its environmental footprint. Developing the Model X was a demanding task that involved several engineering challenges, including the "falcon wing" doors. These doors not only had to look cool, but also offer practical benefits, such as easy access to the rear seats in tight parking spaces.

Design

The Model X stands out with its futuristic yet functional design. The "falcon wing" doors are the most striking design element and allow easier access to the rear of the vehicle. Similar to other Tesla models, the interior is minimalist in design, with a large central touchscreen as the main control interface.

Technical specifications

The Tesla Model X is essentially based on the technology of the Model S. In terms of performance and range, the Model X is therefore not inferior to its siblings. It is offered in various configurations that allow ranges of over 300 miles/550 kilometers and acceleration times from 0 to 60 miles/100 kilometers per hour in about 2.5 seconds. This impressive performance is made possible by the all-wheel drive system and the efficient electric motors.

Other technical features include an adaptive air suspension that can be adjusted for different types of terrain and a trailer hitch with high towing capacity for those who want to carry extra luggage or even a caravan.

The Model X has received generally positive reviews, especially for its performance and technological innovations. However, it also has some criticisms, mainly due to its high price and some teething problems in early production models. Nevertheless, it has fundamentally changed the perception of what an SUV can and should be and pushed the industry toward a more sustainable future.

Overall, the Tesla Model X is a prime example of combining luxury, performance, and sustainability in an SUV that, according to popular opinion, is not really suited for electrification due to weight and drag.

Tesla Model 3

Since its launch in 2017, the Tesla Model 3 has not only shaken up the electric vehicle market, but also challenged the automotive industry in general. With its combination of affordable pricing (base price around 43,000 euros), high range and advanced technology with full integration into the "Tesla world", the Model 3 has made sophisticated electric mobility accessible to a wider mass.

Technical specifications

One of the main selling points of the Tesla Model 3 is its impressive range for its class. Depending on the model, buyers can expect a range of up to 353 miles (about 568 kilometers) on a single charge. That's a significant advantage over many other electric cars on the market, which often struggle to exceed 200 miles. In addition, the car offers impressive acceleration capabilities, with some variants making it from 0 to 60 miles per hour in just 3.1 seconds.

Design and interior

The exterior design of the Model 3 is simple but elegant. The kinship to the Model S is obvious. The interior design is even more minimalist, with a central 15-inch touchscreen that controls almost all the car's functions. There are no traditional buttons or switches; almost everything is controlled by the touchscreen. This creates a clean, modern look and allows drivers to focus on what's important.

Autonomous driving and software

One of the key features of the Tesla Model 3 is its advanced software, which is continuously improved through over-the-air updates. The car's Autopilot features are particularly impressive, including functions like lane change, parking assist, and even a "summon" mode where the car is capable of driving autonomously to its owner.

Economic and environmental impact

The Tesla Model 3 has given the electric car market a much-needed boost. Its relatively affordable prices and high sales figures have increased the pressure on other automakers to become more competitive in the electric segment. In addition, the Model 3 plays a crucial role in the transition to more sustainable transportation and helps reduce greenhouse gas emissions.

The Tesla Model 3 is far more than just another electric car; it's a paradigm shift in the automotive industry. With its combination of range, performance, design, and autonomy, it has not only changed the way we think about electric vehicles, but also how we think about mobility in the 21st century. It has accelerated the mass market introduction of electric vehicles and paved the way for a more sustainable future.

Tesla Model Y

The Tesla Model Y is a compact electric SUV based on the same platform as the Model 3. It was unveiled in March 2019 and is part of Tesla's strategy to reach a broader range of consumers.

Development

The Tesla Model Y was developed as part of Tesla's "S3XY" product line. The goal was to leverage the technological advances and production efficiencies Tesla had achieved with the Model 3 to bring a more affordable yet powerful electric SUV to market. The development process benefited from the experience Tesla had gained with its previous models and aimed to reduce production costs and improve manufacturing quality.

Design

The Model Y's design is a natural evolution of Tesla's design language and shares many aesthetic characteristics with the Model 3, but it offers more cargo space and an optional third row of seats to accommodate up to seven passengers. Similar to other Tesla models, the Model Y features a minimalist interior with a central touchscreen that controls most of the car's functions.

Technical specifications

The Model Y comes in several variants that differ in range, performance, and price. The electric motor provides impressive acceleration, and the battery technology offers a competitive range of over 300 miles, depending on the model variant. The dual-motor all-wheel-drive version improves traction and driving stability, which is especially useful in harsh weather conditions.

Vehicle safety is another key focus, with numerous airbags, a low center of gravity to minimize rollover risk, and advanced driver assistance systems, including Tesla's Autopilot features.

Software and updates

As with the Model S, one of the most innovative features of the Model Y is its ability to receive software updates "over-the-air". This means that the car actually gets better over time, as new features can be added, and existing ones optimized without requiring a visit to the dealer.

Reception and market impact

The Model Y has established itself as a popular model in Tesla's lineup, especially for those looking for a more sustainable alternative to traditional gasoline-powered SUVs. By combining range, performance, and functionality, the Model Y has helped to further drive electric vehicle adoption and encourage competitors to accelerate their own EV offerings.

Overall, the Tesla Model Y has been instrumental in continuing Tesla's mission to move the world faster toward sustainable energy. It has proven that a small electric SUV can not only be feasible, but superior in many aspects, and it has redefined expectations of what is possible in this class of vehicle.

Tesla Cybertruck

When Elon Musk unveiled the Tesla Cybertruck to the public in November 2019, the reaction was mixed but inevitably attention-grabbing. The futuristic, angular design was either a feast for the eyes or a thorn in the side, depending on your perspective. But one thing was clear:

The Cybertruck was unlike anything else in the pickup truck market. The truck is a clear departure from traditional automotive aesthetics and embodies an ambitious vision for the future of mobility.

Design: A look into the future

The design of the Cybertruck is nothing less than revolutionary. With its angular, almost dystopian aesthetics, the vehicle is more reminiscent of a futuristic war vehicle than a conventional pickup truck. Its outer shell is made of cold-rolled stainless steel, which is not only rustproof but also particularly tough. The material is so robust that it can even withstand small-caliber firearms, which was demonstrated during the presentation in a live demo that was not without controversy.

copyright tesla.com

Powertrain and performance

The Cybertruck will be offered in a variety of configurations, from a single-engine rear-wheel-drive model to a three-engine all-wheel-drive version. The most powerful model will be able to accelerate from 0 to 60 miles per hour in just 2.9 seconds. In addition, the highest model is said to offer a range of up to 500 miles, which is well above average for electric vehicles, let alone electric pickups.

Functionality and versatility

The Cybertruck is not only intended to be fast and powerful, but also extremely functional. The cargo area is generously dimensioned and fitted with an electrically operated cover. In addition, features such as an adaptive air suspension system and an integrated ramp extension for easy loading and unloading are planned. The interior seats up to six people and integrates state-of-the-art technology, including a 17-inch touchscreen similar to those already used in the Model S and Model X.

Potential impact on the industry

The cybertruck comes at a time when the debate about sustainable mobility and climate change is more intense than ever. Its introduction could be a turning point for the acceptance of electric vehicles in a segment traditionally dominated by gasoline or diesel-powered models. If the cybertruck delivers on its promise, it could serve as a catalyst for broader acceptance of electric pickups and commercial vehicles.

The Tesla Cybertruck is more than just another electric vehicle; it's a statement, a provocation, and perhaps a glimpse into the future of the automotive industry. Although polarizing, it has the potential to fundamentally change the way we think about mobility, performance, and design. What's certain is that even now, ahead of its official launch, expected in late 2023 (Europe: 2024), it's already creating a lot of buzz and keeping the industry on its toes. Only time will tell whether the cyber truck will have the revolutionary effect that is attributed to it. But one thing is certain: it has already attracted the world's attention, and that is an achievement in itself.

Tesla Roadster (2nd generation)

The supercar of the future: The announcement of the new Tesla Roadster in 2017 was much noticed in the automotive world. Now planned for a 2024 launch, the Roadster promises to radically push the boundaries of what has previously been expected from electric vehicles. From incredible speed to breathtaking range, the Tesla Roadster stands as a symbol of the next era of electric mobility.

Design: Aesthetics meets aerodynamics

The design of the Tesla Roadster is a blend of elegance and functionality. The aerodynamic shape of the vehicle is not only aesthetically pleasing, but also functional, as it further minimizes air resistance. One feature is the removable targa roof. With its sleek lines and curves, the vehicle seems to be literally shaped for speed.

Powertrain: Next-generation performance

Equipped with three electric motors - two on the rear axle and one on the front - the Roadster aims to take a barely comprehensible 1.9 seconds to sprint from 0 to 60 miles/100 kilometers per hour. With a top speed of over 250 miles/400 kilometers per hour, it would not only be the fastest electric car, but could easily compete with the very fastest gasoline-powered supercars. Even the fastest Bugatti would reach its limits.

Reach: A new paradigm

One aspect of electric vehicles that is often criticized is their limited range. The Tesla Roadster aims to finally eliminate this prejudice with a planned range of 620 miles on a single charge. This corresponds to a range of around 1,000 kilometers. Such a breakthrough would only be possible with an enormous battery capacity, and Tesla plans to achieve this goal with a 200-kWh battery pack that is twice the size of anything currently available in the Tesla fleet.

Interior and technology

Inside, the roadster promises an equally futuristic experience. The minimalist interior is said to feature four seats and a central touchscreen display. Although details about the specific technology used in the Roadster are still sparse, it is likely that it will not only benefit from Tesla's extensive experience in autonomous driving features but push their boundaries further.

Significance for the automotive industry

The Tesla Roadster has the potential to be more than just another luxury electric car. It could become a catalyst for the entire industry, demonstrating that electric vehicles can not only match but surpass their gasoline-powered counterparts in terms of performance, range, and design. If the Roadster delivers on its promise, it could boost acceptance of electric vehicles in segments of the market that have been skeptical of the technology.

The second-generation Tesla Roadster stands as an iconic example of the rapid development of electric mobility. It embodies the ambition and vision that have driven Tesla since its inception. It offers a radical reinterpretation of what is possible in the world of electric vehicles - a vision of the future on four wheels.

Tesla Semi (Truck)

Since its announcement in 2017, the Tesla Semi (= semi-truck) has attracted some attention both within the transportation industry and the broader public. The first vehicle was delivered in the U.S. in late 2022, significantly behind schedule. As an electric truck that has the potential to fundamentally change freight transportation, the Tesla Semi is nevertheless at the forefront of a wave of innovation that could one day revolutionize the entire road transportation industry.

Technical specifications and performance

The Tesla Semi will come in two main versions that differ substantially in range: one with 300 miles or 500 kilometers and one with 500 miles or 800 kilometers of range per charge. The truck will be powered by four independent electric motors distributed among the rear axles. These motors are variations of those used in the Tesla Model 3 and promise high efficiency and impressive performance figures. The semi is said to be able to accelerate from 0 to 60 miles/100 kilometers per hour in just 5 seconds without a load and in about 20 seconds with a full 80,000-pound load (about 36 tons).

In terms of charging technology, Tesla plans to create a network of "megachargers" that will enable rapid charging. Within 30 minutes, these charging stations could provide enough energy for 400 miles (650 kilometers) of new range.

Design and aerodynamics

One of the standout features of the Tesla Semi is its futuristic design. With an aerodynamic shape that minimizes drag, the design aims to reduce energy consumption and thus increase range. The driver's area is designed to provide good visibility and control, and the interior is equipped with state-of-the-art technology, including multiple touchscreens for navigation, monitoring and system control.

Economic and environmental importance

The economic impact of the Tesla Semi could be significant in the future. By transitioning from diesel to electric, transportation companies could realize significant cost savings in fuel and maintenance. In addition, the truck has the potential to significantly reduce emissions in the freight sector, one of the largest emitters of greenhouse gases.

Similarly, the introduction of the Semi could mark a turning point for the acceptance of electric vehicles in an industry segment that has traditionally been reluctant to embrace new technologies. The Semi's success could lead to other manufacturers developing similar electric truck models, which would accelerate the transition to more sustainable transportation solutions.

The Tesla Semi is more than just another step in the evolution of electric vehicles; it has the potential to be a paradigm shift in the way goods are transported. From its impressive performance to its advanced technology and sustainable design, the Semi promises to fundamentally change the transportation industry. Although the vehicle is not yet in mass operation and many technical and economic questions remain unanswered, all indications are that the Tesla Semi could be that disruptive factor in an industry that desperately needs sustainable alternatives.

What kind of vehicle is Tesla missing?

Tesla has launched no small range of electric vehicles in recent years, ranging from luxury sedans to an electric truck. Still, there are some vehicle categories in which Tesla has not yet been active.

Despite the success of the "mass" Model 3 as a relatively affordable electric vehicle, Tesla still lacks a true compact or city car. Such a vehicle could be ideal for urban environments and further lower the barrier to entry for electric cars. It would also allow Tesla to better compete in markets like Europe or Asia, where smaller cars are often preferred.

Another category currently missing from the Tesla lineup is the minivan or family van. Such vehicles could be a way for Tesla to appeal to a broader range of buyers.

Although Tesla is entering the commercial vehicle market with the Tesla Semi, there are no known plans yet for smaller vans or delivery trucks. These could be of great interest to small businesses or tradesmen and would lead Tesla into a market that will be increasingly affected by electrification, as delivery traffic often takes place in inner cities. However, competition is also particularly high in this sector.

Tesla has so far focused entirely on four-wheeled vehicles. However, the market for electric motorcycles and scooters is also growing, and a Tesla two-wheeler could be quite interesting there.

Apart from the Cybertruck, which will be a rather futuristic interpretation of a pickup truck, Tesla also does not have a true off-road vehicle in its lineup. Such a vehicle could be popular with a target group that appreciates adventure and off-road driving.

So there are several vehicle categories where Tesla does not yet have a presence that the company could consider to increase its market reach and depth.

How are Teslas distributed?

The distribution of Tesla vehicles is fundamentally different from that of traditional car manufacturers. Instead of relying on a dealer network that operates independently of the brand, Tesla has established a direct sales model that brings the manufacturer closer to the end consumer. This model has brought both benefits and challenges that could impact the automotive industry in general.

A prominent feature of the Tesla sales model is online sales. Customers buy a Tesla car directly on the company's website. The entire purchasing process, from color and equipment selection to financing, is to be handled online. This digital approach minimizes the need for physical sales outlets and fits well with Tesla's tech-savvy target group, but it also prevents dealers from giving discounts, for example. At any rate, new Teslas can currently only be purchased directly from the manufacturer, who generally does not offer discounts on the final price. The competition between dealers that otherwise occurs is missing here.

Although online sales are the focus, Tesla also operates showrooms and so-called "Experience Centers" in major metropolitan areas. These physical locations serve several purposes: they are showcases for the brand, places for test drives, and sources of information for potential buyers. However, they are not traditional car dealerships. As a rule, no transactions are conducted here; instead, customers are encouraged to buy online.

Tesla's direct sales model has faced legal challenges in some states in the U.S. and in some other countries. Traditional car dealers and their lobbying associations have argued against Tesla's sales methods, often citing existing laws that prohibit or restrict direct sales of cars by manufacturers. Tesla has met these challenges in a variety of ways, including lobbying for legal changes or using exemptions.

Typically, such direct sales models in the automotive market can only be enforced by strong brands, while weaker ones consistently distribute through dealers. Tesla's situation is similar to that of Mercedes-Benz in the 1980s, when new vehicles were mostly distributed rather than sold without any discount opportunity. This has changed.

Typically, such a distribution system based on high demand is accompanied by longer delivery times; however, it seems that Tesla has this largely under control, at least in most countries and markets.

In the global context, Tesla has expanded its presence by building gigafactories and distribution centers. Countries such as China have been targeted both as production locations and as important sales markets. Global sales in this case are also mainly via the direct sales model.

Another important element of Tesla sales is after-sales service. After-sales services such as maintenance and repair are also provided by mobile service teams. The company also uses its software expertise to provide over-the-air updates for the vehicles, minimizing the need for physical service visits.

Overall, Tesla's sales approach has maximized the benefits of both digitization and a direct customer relationship. However, this model also challenges the existing practices and legal framework of the automotive industry and, for this very reason, could serve as a blueprint for other strong manufacturers in the future.

How are Teslas repaired?

Repairing a Tesla vehicle differs in some respects from repairing conventional internal combustion vehicles, as Tesla uses a number of special technologies and systems that are not common property.

Most repairs are carried out at Tesla Service Centers. These are specifically geared to the needs of Tesla vehicles and have trained personnel and special tools. For many types of repairs, especially those involving the vehicle's electronics or software, there is no alternative anyway.

One of the largely unique aspects of Tesla vehicles is the ability to receive or install software updates over-the-air (OTA). In some cases, problems or bugs can be fixed with a simple online software update. Tesla also has the ability to perform remote diagnostics to check a vehicle's condition and identify any problems. Further, Tesla offers the ability to book maintenance work online via the app, for example.

There are also a few independent repair shops that specialize in repairing Tesla vehicles. Their numbers are likely to increase over time as more Teslas get older, come out of warranty and the need for cheaper repairs increases.

Another issue may be that Tesla controls the distribution of its replacement parts, which means that even independent repair shops often must purchase these parts directly from Tesla. This can have a negative impact on the cost and availability of third-party repair services but may mitigate in some markets as demand for parts increases. In addition, in some markets, such as within the EU, there is the phenomenon that strong market manufacturers may be required to provide parts, information and, for example, software to third parties outside their own distribution network. In practice, however, this often fails because of the costs involved.

Can independent workshops handle Teslas?

The ability of independent or chain repair shops to repair Tesla vehicles is a complex issue.

Not all independent repair shops have the specialized knowledge and equipment needed to repair Tesla vehicles. Tesla vehicles contain complex electrical systems and working on the high-voltage battery or the electric drive components requires special training and tools. In addition, independent repair shops generally do not have Tesla-specific software. This lack of competitive pressure can also mean that maintenance and repairs tend to be relatively expensive for Tesla as a strong market player.

Teslas are largely software-driven, and many repairs or diagnostic procedures require access to specialized diagnostic tools and software offered only by Tesla or by Tesla-certified repair shops. However, some independent repair shops have found ways to gain access to some of these tools or use alternative diagnostic methods.

In some countries, such as the European Union, there are efforts to give legal support to the "right to repair" movements, which would make it easier for independent repair shops to access necessary diagnostic tools and spare parts. This could affect the ability of independent repair shops to repair Teslas in the future. Currently, it is not a major issue because of the relatively young age of the Tesla fleet, but this may change in the future. Another factor here is that Teslas, more than vehicles from other manufacturers, are a largely closed system of hardware, software and masses of data that is difficult to intervene in selectively without taking risks elsewhere.

Overall, independent repair shops can repair Teslas, but the challenges of expertise, parts sourcing, and software access mean that this is currently very limited in practical terms.

Where to get spare parts for Teslas

Acquiring replacement parts for Tesla vehicles can be different than traditional car brands in some areas, and there are currently fewer ways to obtain these parts.

The most obvious way to get spare parts for a Tesla is through Tesla Service Centers. These facilities specialize in servicing and repairing Tesla vehicles, and they have direct access to genuine spare parts. The downside could be higher costs, as the competition that otherwise exists from independent dealers is still largely absent.

There are a number of online dealers and third-party suppliers that sell replacement parts for Tesla vehicles. These may offer used parts from disassembled vehicles or replica parts from other manufacturers. Although these parts are often less expensive, the quality varies, and they often come without a warranty. In addition, using non-original parts can affect the vehicle's manufacturer's warranty, which of course is not unique to Tesla. Some independent repair shops that specialize in Tesla or electric vehicles may also have access to replacement parts.

The Tesla community is quite active, by the way, and in various forums and online platforms you can often find useful information about where to find specific parts or how to perform certain repairs yourself.

Batteries and charging network

Tesla has established itself as a leader in electric vehicles and energy storage, and part of that success is due to its advanced battery technology.

Tesla has developed a special lithium-ion cell chemistry that enables high energy density. This means that the batteries can store more energy without significantly increasing the size or weight of the battery. Tesla originally used an 18650-cell shape (18 mm in diameter and 65 mm long), but has since switched to the larger 2170 format (21 mm in diameter and 70 mm long), which offers even higher energy density. Tesla is also working on developing new cell formats, such as the "4680" cells, which promise even more capacity and power.

Tesla's battery systems are modular. This means they consist of many individual cells that are combined in modules and packs. This scalability enables a wide range of applications, from the smaller battery packs in Tesla vehicles to the large Tesla megapacks for industrial applications.

Tesla's battery systems have sophisticated temperature management systems that keep the cells at optimal temperatures. This is critical for battery performance, longevity, and safety. Temperatures that are too high or too low can affect performance and shorten cell life.

The ability to be charged quickly is another major advantage of Tesla batteries. The fast-charging capability of Tesla's batteries is the result of a combination of factors, including the chemistry of the lithium-ion cells used, the battery management system, software optimizations and the charging infrastructure itself. Tesla uses several types of lithium-ion batteries, but they are all designed to support fast charging cycles with high efficiency. For example, some models use NCA (nickel-cobalt aluminum) cells, which provide a good balance between energy density and thermal stability.

Tesla's integrated battery management system plays a crucial role in monitoring and controlling the charging processes. It regulates the voltage and current flow to ensure that the batteries are charged safely and efficiently. It also uses algorithms to determine the optimal charging cycle for each driving situation and battery condition.

Tesla invests heavily in research and development of software algorithms that make charging more efficient. Over-the-air updates allow them to regularly improve these algorithms and distribute them to drivers. These tweaks can improve the battery's fast-charging capability over time.

Tesla has developed its own network of Supercharger charging stations specifically designed to charge Teslas as quickly as possible. These Superchargers are capable of delivering very high charging power (up to 250 kW for some models), which significantly reduces charging times.

The batteries are cooled during the charging process by a sophisticated thermal management system. This system ensures that the battery remains within an optimal temperature range, which increases both charging efficiency and battery life.

An often overlooked but critical aspect of Tesla's battery technology is the software that controls the battery management system (BMS). This system monitors the battery's performance, regulates temperature, and maintains an even load on the cells, which extends its life. Tesla's BMS is also capable of receiving over-the-air updates, so the battery's performance and functionality can be improved over time. This also applies to the regeneration of older batteries.

Tesla's battery technology is not just limited to vehicles. It is also an integral part of other Tesla products such as the Powerwalls and Megapacks for energy storage, which opens synergies and new areas of application.

Tesla has focused on researching and developing more sustainable battery manufacturing methods, including recyclability, and reducing the use of rare materials such as cobalt.

Battery replacement and recycling

The batteries in Tesla vehicles can be replaced, but such replacement is usually complex and expensive.

The battery life in Tesla vehicles is typically long and covered by a limited warranty. However, battery replacement might be necessary due to a defect, an accident, or a significant reduction in capacity over time. Tesla has dedicated service centers equipped with the necessary knowledge and specialized tools for such a replacement. Depending on the model and capacity, battery replacement can be very expensive and may exceed the current value of the vehicle.

The battery in a Tesla vehicle is more than just an energy storage device; it is part of a complex system that also requires tight software integration. After replacement, appropriate adjustments must be made in the vehicle software to ensure optimal performance and safety. This is another reason why such a replacement should be carried out by specialized technicians.

Tesla has programs to recycle used batteries. These batteries contain valuable materials that can be recycled and reused in new batteries.

For some older Tesla models, like the Tesla Roadster (2008), the company now offers special battery upgrades that can improve range and performance.

In recent years, the first independent providers have established themselves that either replace or repair Tesla car batteries in order to carry out repairs that are in line with the current value. It is to be expected that such providers will increase significantly, so that repair options can often be found for older Teslas as well. It also plays a role here that older Teslas, more than many other vehicles, also have the character of a long-term car whose value is primarily based on the battery condition. Therefore, more expensive repairs to the battery storage can also be worthwhile here.

The Tesla charging network

The Tesla Supercharger network is a key element in Tesla's success as an electric vehicle manufacturer and offers some unique benefits. Tesla's Supercharger stations are known for offering fast charging speeds. Depending on the model and charging station, Tesla vehicles can charge up to 80% of their capacity in just 20 to 30 minutes. That's much faster than most other publicly available charging options and makes long-distance driving in an electric vehicle more feasible. Other manufacturers don't yet offer that in this form.

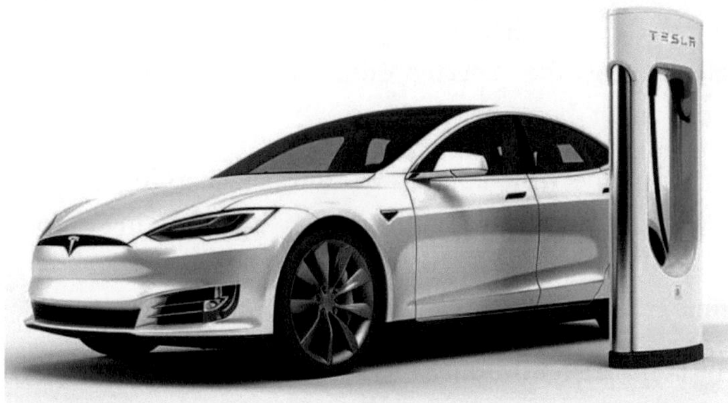

The Superchargers are currently still designed and optimized exclusively for Tesla vehicles. This ensures that the stations are usually available when Tesla owners need them and eliminates compatibility issues that can occur with other charging infrastructure. However, Tesla is in the process of opening its charging infrastructure to competitors.

Tesla has established an extensive network of Superchargers in many countries, which is constantly being expanded. This gives Tesla drivers the freedom to drive long distances without range anxiety. The stations are strategically placed to cover major traffic arteries and long-distance routes.

Charging at a Supercharger is very user-friendly. No separate subscription or card is required; everything is handled via the owner's in-car navigation system and Tesla account. The car communicates directly with the charging station, and billing is automatic.

Tesla's vehicle software is closely integrated with the Supercharger network. For example, the navigation system takes into account the battery's state of charge and automatically plans charging stops on long-distance trips. It can also forecast the expected utilization of Supercharger stations along the route.

Through the Tesla app, users can track charging status in real time and receive notifications when charging is complete. This increases convenience and allows drivers to use time efficiently during the charging process.

Tesla has also introduced the so-called V3 Superchargers, which offer even higher performance and faster charging times. In addition, the company has announced plans to further improve and expand the charging infrastructure, both in terms of the number of stations and the technology itself.

Some Supercharger stations are already equipped with solar panels, promoting the use of renewable energy. This fits with Tesla's overall goal of promoting sustainable energy solutions.

In addition to the classic Superchargers, which are often located on highways and main traffic arteries, Tesla has also introduced "Urban Superchargers." These are found in urban areas and near shopping centers and offer slightly reduced charging power but are ideal for people who want to charge quickly in urban areas.

The Supercharger network uses intelligent software for dynamic load management. This means that the available electrical power is efficiently distributed to the vehicles that are being charged. For example, when one vehicle is almost fully charged, more power can be distributed to other, less charged vehicles that can make better use of the extra power.

In some areas, Tesla is experimenting with amenities around the Superchargers such as comfortable waiting areas, Wi-Fi, snacks and even cafés. These extras are intended to make charging time more pleasant and thus promote the acceptance of electric vehicles.

Some Supercharger stations are equipped with their own battery storage (usually Tesla Powerpacks or Megapacks) to better manage power demand and provide reliable power in emergencies.

While use of the Superchargers was originally free, Tesla has since introduced a paid model. Nevertheless, the Supercharger network remains economically attractive compared to many other fast-charging options. Tesla also offers occasional promotions, awarding free Supercharger miles for referrals or the purchase of certain models.

Tesla has a global network and is striving to expand this even further. Superchargers, for example, are not only widespread in North America and Europe, but also in countries such as China and Australia.

The combination of all these factors results in a robust, reliable, and user-friendly charging infrastructure that is an essential part of the Tesla experience and goes a long way toward promoting electric mobility. It's not just the technology itself, but the thoughtful integration across the company's ecosystem that makes the Tesla Supercharger network so special.

Battery storage

In addition to cars, Tesla also offers battery storage systems designed for residential, commercial, and industrial use. These systems are separate from the batteries in Tesla vehicles and serve the purpose of storing electrical energy that can then be used at a later time.

The Tesla Powerwall is a battery storage system for private use and is often installed in conjunction with a photovoltaic system (solar panels). The Powerwall can store energy when the sun is shining and release that energy when it is needed, such as at night or during a power outage. This allows homeowners to maximize their self-consumption of solar power and reduce their dependence on the power grid. The system can also be used to store power from the grid at times of lower demand and then release it at peak times, which can result in cost savings in some areas.

While the Powerwall is designed for domestic use, the Tesla Powerpack and Tesla Megapack are designed for commercial and industrial applications. These larger storage systems have significantly higher capacities and are designed to be connected to the power grid to regulate energy flow, cover peak loads, or even act as independent energy sources. For example, these systems can be used to store solar or wind energy and then release it when demand is high, or power generation is low.

Again, an important aspect of Tesla battery storage is the software that controls it. Tesla's energy management software allows intelligent control of energy flow to maximize self-consumption, minimize costs, and even feed energy back into the grid if allowed.

Digitization to an end

Digitization plays a central role in Tesla's strategy and products.

One of the most revolutionary aspects of Tesla's approach to begin with is the ability to push out software updates "over-the-air" (OTA). This allows Tesla to not only fix small bugs, but also add significant improvements and even entirely new features. This was a significant advance over traditional automakers, where software updates typically required a visit to the repair shop, although others do it similarly today.

Tesla develops most of the software and hardware for its vehicles itself. This enables seamless integration and efficiency that is difficult to achieve with third-party solutions. From the electric powertrain to the user interface of the infotainment system, everything is coordinated.

Tesla is a pioneer in autonomous driving systems with its "Autopilot" and more advanced "Full Self-Driving" (FSD) options. The system is continuously improved through the constant collection of data and machine learning algorithms.

Tesla's vehicles constantly collect data that is used to improve the software and hardware. By analyzing this data,

Tesla can, for example, improve the behavior of Autopilot, optimize battery life, or even make predictive maintenance forecasts.

Tesla cars are constantly connected to the internet, which not only enables OTA updates, but also features like real-time traffic data, streaming services, and even video games right in the car.

The infotainment system, which is usually controlled by a large touchscreen in the center of the dashboard. It offers a wide range of functions, from navigation to climate control, and it is constantly being developed further.

Digitization also plays a role in the efficient energy management of Tesla vehicles. From intelligent route planning that considers the battery's state of charge to sophisticated algorithms that minimize energy consumption, software makes a significant contribution to the vehicles' performance.

The buying experience of a Tesla also starts mostly digitally. From online configuration of the car to delivery, Tesla relies on digital interfaces. There are no traditional dealerships; instead, the buying process is mostly online.

The Tesla Mobile App is another example of deep digitization. Through the app, owners can access many of the vehicle's functions, including turning the air conditioning on and off, monitoring the battery's charge status, locating the parked vehicle, and even summoning the vehicle from a parking spot.

By using the digital data collected from its vehicles, Tesla is able to make continuous improvements that go far beyond fixing bugs. The company uses artificial intelligence and machine learning to continuously optimize Autopilot, for example.

Tesla connects its cars to a broader digital ecosystem. In addition to the car itself, other Tesla products, such as solar panels and home batteries, are connected and can be managed via the same app.

The strongly software-centric architecture allows Tesla to respond very flexibly to changes in the market or in technology. Whereas hardware upgrades in conventional cars are often only possible with the purchase of a new model, digitization at Tesla enables the continuous adaptation and improvement of existing models, which also benefits their service life.

Cybersecurity, data protection

In an era when cars are increasingly becoming networked computers, cybersecurity is becoming more and more important.

Tesla is a company that operates in the automotive, energy storage, and artificial intelligence industries, and cybersecurity is critical for such technology companies. Tesla has taken a number of measures to strengthen cybersecurity, but like any other company, Tesla is not completely immune to cyber threats.

Tesla vehicles are designed to receive over-the-air (OTA) updates. This allows the company to quickly close security gaps. Once a vulnerability is identified, a patch can be sent to all affected vehicles in near real-time. However, vehicle-specific gateways need to be specially monitored, as they can themselves be a vulnerability.

Tesla operates an active bug bounty program that allows independent security researchers to identify potential security issues and receive rewards for doing so. This helps Tesla discover potential vulnerabilities before they can be exploited by malicious actors.

Encryption is another important element of Tesla's strategy. The company uses advanced encryption mechanisms to protect data communications between the vehicle and the company's servers.

Tesla also emphasizes the physical security of its vehicles. Features such as camera surveillance and alarm systems are examples of how Tesla increases security on a physical level.

Tesla integrates cybersecurity into the product development process from the beginning. This includes both hardware and software elements to ensure that the entire system architecture is robust against potential attacks.

Despite these efforts, Tesla has also experienced some prominent cybersecurity incidents in the past that have prompted the company to continually revise and improve its security measures.

Tesla vehicles and products generate numerous data points that can be used for a variety of purposes. From vehicle diagnostics to improving autonomous driving capabilities and providing personalized services, Tesla collects a wide variety of data.

Tesla has a publicly available privacy policy that explains what types of data are collected and how they are used. These policies are established in accordance with legal requirements and are updated regularly.

Tesla strives to minimize data, i.e. to collect and store only the data that is really necessary. The company also states that it stores data only as long as necessary and deletes or anonymizes it when it is no longer needed.

Tesla offers users a few ways to control their data. Through the user account and, in some cases, directly in the vehicle, users can adjust their privacy settings and specify what data they want to share.

Tesla complies with the data privacy laws of the countries in which it operates. This includes, for example, the European General Data Protection Regulation (GDPR) and the California Consumer Privacy Act (CCPA) in the USA.

In addition to encryption, Tesla also takes internal security measures to prevent unauthorized access to data. This includes technological solutions such as firewalls and security-checked network infrastructures as well as organizational measures such as training and security audits.

Tesla also undergoes external third-party audits and maintains industry-standard certifications to demonstrate compliance with privacy standards.

Tesla's Supercharger network and service centers are also digitally integrated. The cars know where the next charging station or service point is and can even plan the optimal route, considering the current battery level.

The Tesla app

The Tesla app, in turn, is an integral part of the Tesla experience and serves as a digital interface between the vehicle and its owner.

The app allows users to check the status of their vehicle in real time. This includes information such as the battery charge level, range, location of the vehicle and much more. They can also lock and unlock the doors, turn the air conditioning on or off, and even open or close the windows.

One of the features of the app is the ability to control the charging process. Users can start or stop charging, schedule a charge time, and even set the desired charge limit to optimize battery health.

For models equipped with these functions, the app allows you to "call" the vehicle from a distance or have it park automatically. This is particularly useful in tight parking spaces or when loading heavy shopping into the car.

Destinations can be sent directly to the Tesla's navigation system via the app. In addition, you can view the current location of the vehicle at any time, which is especially useful if you have parked your car in a large parking lot or in unfamiliar areas.

The ability to control the air conditioning or heating remotely is another useful feature. You can preheat or cool the car before you get in, which is especially handy in extreme weather conditions.

The app also informs the owner about upcoming or performed software updates. In addition, it can send notifications, for example, when the charging process is complete or when a door has not been properly locked.

For those who also own other Tesla products, such as solar panels or powerwalls, the app provides an integrated platform for monitoring and controlling these systems.

Finally, the app also makes it easier to schedule service appointments. You can book an appointment directly in the app without having to make a call or visit a separate website.

The Tesla app is designed to be intuitive and user-friendly, and it is regularly updated to introduce new features and improvements. These comprehensive features make the app a central component in the Tesla ecosystem and greatly increase convenience and control for owners.

What makes the Tesla app special is its comprehensive integration into the entire Tesla ecosystem, providing a seamless and intuitive user experience. While many car manufacturers offer apps with limited features, Tesla goes far beyond that, making the app a kind of remote control for the car and even for other Tesla products like solar panels and energy storage.

The app is seamlessly integrated with Tesla's vehicles and products. This means that users have a unified and consistent experience whether they are driving the car themselves, using the app, or operating other Tesla products.

The app enables real-time interaction with the vehicle. You can see the car's current location, check the battery's charge status, and even preheat or cool the interior before you get in.

The app is known for its simple and intuitive user interface, which is easy to understand even for people who are not particularly tech-savvy. The design is clean and straightforward, so users can quickly find the information and features they need.

The autopilot

Tesla Autopilot is one of the best-known and most discussed assistance systems for autonomous driving on the market. It is important to emphasize that despite its connotations, the name "Autopilot" does not mean that the vehicle can drive fully autonomously. Instead, it is an advanced driving assistance system that operates under the supervision of a human driver.

Elements of the autopilot

One of the basic functions of Autopilot is adaptive cruise control, which enables the car to automatically adjust its speed to the flow of traffic. The system can reduce speed when following a slower vehicle and accelerate again when the road is clear.

The lane departure warning system is another basic component. This function ensures that the vehicle stays within its lane. To do this, Autopilot uses cameras and sensors to detect lane markings.

This advanced feature allows the vehicle to take freeway exits, change lanes, and even select the correct exit within an interchange to follow a preset navigation destination.

"Summon" allows the vehicle to drive in and out of parking spaces or garages without a driver in the car. "Smart Summon" is an enhanced version of this feature and allows the vehicle to drive to its owner in a parking space, provided the owner is in sight.

Newer versions of Autopilot can recognize traffic signs and lights and adjust the speed or behavior of the vehicle accordingly.

One of the most innovative aspects of Tesla Autopilot is its "fleet learning" capability. Since all Tesla vehicles are constantly collecting data, the company can use this information to continuously improve the system.

While Autopilot is active, the driver must still have his hands on the steering wheel and be ready to take control. The system uses interior cameras and sensors in the steering wheel to ensure the driver is alert.

Tesla offers an additional option called "Full Self-Driving" or FSD, which adds several advanced features to the basic Autopilot system. However, it is important to emphasize that despite the name, even FSD is not fully autonomous and requires human supervision. Additional FSD features include the ability to automatically change lanes, stop at intersections and stop signs, and even navigate autonomously through urban streets.

The use of Tesla Autopilot and similar systems naturally raises ethical and legal questions, especially in the event of accidents. Who is responsible if something goes wrong? These questions are the subject of ongoing debates and regulatory considerations.

The Tesla Autopilot system relies on a variety of sensors, including cameras, radar, and ultrasound, as well as advanced algorithms for data processing and machine learning. This technological infrastructure is critical to the system's accuracy and reliability.

Since Tesla vehicles are electric, Autopilot has also been optimized for energy efficiency. The system is designed to keep battery consumption as low as possible without compromising performance.

Limits and problems of the autopilot

Tesla Autopilot is a revolutionary technological system that can make driving safer and more comfortable, but it is not without its challenges and criticisms.

The name "Autopilot" can be misleading and lead users to believe that the system can drive fully autonomously. This can lead to overuse of the functions and reduced driver attention, creating potentially dangerous situations.

Another problem is what is known as "monitoring fatigue."
When a system is working well most of the time, people
tend to lose their attention, which can lead to delays in human intervention if the system suddenly fails.

Although autopilot is very advanced, it has technical limi-
tations. For example, the system has difficulty operating in
certain weather conditions such as heavy rain or fog. In addition, there can be problems recognizing stationary ob-
jects or distinguishing between a range of traffic signs and
signals.

The increasing proliferation of autopilot systems raises legal and ethical questions, especially in the event of an accident. Who is responsible? The driver, the system or the
manufacturer? These questions have not yet been conclu-
sively clarified and are the subject of ongoing investiga-
tions and legal proceedings.

Regulation of autonomous and semi-autonomous driving
systems varies from country to country, making the development and implementation of such systems complicated.
In some jurisdictions, certain autopilot functions are re-
stricted or not allowed at all.

The cost of purchasing the full Autopilot or Full Self-Driving package is also high.

There are no industry-standard tests or certifications for
autonomous driving systems, making it difficult to evalu-
ate the safety and effectiveness of Tesla Autopilot compared to similar systems from other manufacturers.

Because the system is constantly being updated, there are
also concerns about long-term reliability and the possibility

that future software updates could cause unforeseen problems or incompatibilities.

Overall, however, Tesla's Autopilot is positive for the following reasons:

Autopilot offers a range of safety functions such as adaptive cruise control, lane departure warning and traffic sign recognition that make driving much safer. The automatic driving assistance functions can make long journeys or the daily commute more comfortable and safer.

Other assistance systems

Tesla has developed a range of assistance systems beyond the basic Autopilot and Full Self-Driving (FSD) options. These systems are designed to make driving safer, more efficient and more comfortable.

For example, there is an advanced version of conventional cruise control that adjusts the vehicle's speed to the flow of traffic. The system uses sensors and cameras to measure the distance to vehicles ahead and adjust the speed accordingly.

Lane Keeping Assist helps the driver stay in the center of the lane. If the vehicle crosses the lines without signaling, Lane Keeping Assist makes corrections to keep the car in the lane.

The smart Summon feature allows Tesla owners to summon their car from a distance via app. It is particularly useful in parking lots, where the car can drive itself to the location where the owner is.

Navigate on Autopilot is an advanced feature of the FSD package that provides active guidance from freeway on-ramp to exit, including lane changes and passing maneuvers.

The Autopark feature enables the car to park itself in parallel or perpendicular parking spaces. All the driver has to do is activate the appropriate mode and the system does the rest.

Sentry Mode is a surveillance system that monitors the car and its surroundings when it is parked. In case of a detected threat, such as someone approaching the vehicle, the system starts recording and can even notify the owner.

The Emergency Lane Departure Avoidance system is designed to prevent accidents caused by unintentional lane departure. The system intervenes when it detects that the vehicle is leaving the lane without signaling and there is a risk of collision.

The Tesla's various cameras can also be used as a dashcam to continuously record video while driving.

Conclusion and outlook

What makes Tesla so fascinating?

- Tesla has fundamentally changed the perception of electric vehicles. While electric cars have long been viewed as slow and impractical, Tesla has developed high-performance vehicles that in many cases outperform conventional gasoline-powered vehicles.

- Tesla is investing heavily in the development of technologies for autonomous driving. Although fully autonomous driving has not yet been realized, Tesla's vehicles already offer a range of "Autopilot" features designed to make driving safer and more comfortable.

- Tesla is not only a carmaker, but also a technology company. With products like the Tesla Powerwall and the Tesla Solar Roof, the company has developed innovative solutions for the energy transition.

- Tesla is breaking with many traditions of the automotive industry. These include direct sales without dealers and a focus on software upgrades delivered over-the-air. This allows the company to make continuous improvements to its products, even after they have been sold.

- Under the leadership of CEO Elon Musk, Tesla has built a tremendous media presence and a strong brand image.

- Tesla has an almost cult-like following, both among consumers and investors. Many people identify strongly with the company's goals of promoting sustainable energy.

- Tesla has shown itself willing to take big risks time and time again. Be it the decision to build a Gigafactory for battery production or the transition from a luxury car manufacturer to a mass producer with the Model 3 - the company's ability to take and manage risks is remarkable.

The appeal of Tesla vehicles themselves can be attributed to several factors, both technological and conceptual.

- The design of Tesla vehicles is unique due to several factors that are reflected in the way they are designed:
- Teslas are electric vehicles (EVs) and the design reflects this electric identity. They don't have grilles because they don't require a traditional internal combustion engine. This allows for a sleeker, futuristic appearance.
- Tesla places great emphasis on aerodynamics to maximize the range of its vehicles. The body shapes are streamlined, which leads to a better drag coefficient and thus improves efficiency.
- Teslas' interior design is minimalist and uncluttered. The dominant feature is the large touchscreen, which serves as the central control element and makes the dashboard look uncluttered.
- Tesla uses high-quality materials in the interior to create a luxurious interior.
- Sustainability: Tesla's design also takes sustainability into account. The use of recycled materials and the focus on energy efficiency help minimize environmental impact.

- One of the biggest barriers to electric vehicle adoption is range anxiety. Tesla has addressed this problem by developing batteries with high capacity and efficiency. Models like the Tesla Model S and Model 3 offer ranges that rival many gasoline-powered vehicles.
- Tesla has created one of the fastest and most widespread charging infrastructures for electric vehicles with its Supercharger network. This makes long-distance travel with a Tesla vehicle largely hassle-free and eliminates another disadvantage often associated with electric vehicles.
- Tesla vehicles are known for their acceleration and good handling. The instant torque of electric motors enables very fast acceleration, and the vehicles' low center of gravity provides excellent driving characteristics that until recently were reserved for sports cars.
- Tesla is a pioneer in the field of driver assistance systems and autonomous driving. Autopilot and the "Full Self-Driving" package offer a range of functions that make driving safer and more comfortable.
- Tesla vehicles offer a minimalist but highly functional interior design with a central touchscreen that serves as the main control for almost all of the vehicle's functions.
- Tesla's brand image as a pioneer and innovator in sustainable energy and high technology attracts many people. In addition, there is a

dedicated community of owners and fans who build strong brand loyalty.
- Tesla places great emphasis on sustainable production and the use of renewable energy, which is an added incentive for environmentally conscious buyers.
- Teslas are exceptionally durable, both in terms of the timeless design, the materials used, and the high-quality technical hardware. The Model S is already in its 13th year without looking old.
- Teslas are considered very stable in value among electric vehicles.

What will happen next?

Tesla has changed, even revolutionized, the automotive industry. Who can revolutionize Tesla for once? It's not the usual suspects like the big competitors in the automotive industry. They must be careful not to be swallowed up by Tesla.

Potential competition is more likely to come from companies like Google or Apple once the vision of autonomous driving becomes reality. And in all likelihood, that won't be too much longer. But Tesla could have the answer to that, too.

One can be curious.

————